5 Strategies for Social Media Engagement

RJ Thesman

Copyright © 2017 RJ Thesman

All rights reserved.

ISBN-13:
978-1976245787

ISBN-10:
1976245788

No part of this publication may be reproduced, stored in a retrieval system, or transmitted in any form or by any means – electronic, mechanical, photocopy, recording or any other – except for brief quotations in printed reviews, without the prior permission of the author.

Printed in the United States of America

5 Strategies for Social Media Engagement

No matter what type of business we are involved in – writing, homespun crafts or farm equipment – social media is now a major part of our lives. To be effective in any type of business, we need to engage with others on some type of social media platform.

Being an effective writer is more than a hobby or a much-loved craft, more than a creative journey into a make-believe world. Writing is also a business. So we have to pay attention to social media trends.

Anybody can post a photo of a scrumptious meal, a selfie with the BFF or those darling pet pictures we all love. But to engage effectively on social media, we need to understand what our audience wants and how we can give it to them.

So let's begin with **Best Practices for Social Media Engagement**:

Know Your Audience – As a writer, I know my audience includes people who love to read. Some of the people in my audience want to read inspirational posts while others are more interested in the mechanics of becoming a writer. Another section of my audience loves funny cat pictures.

When I post, I want to reach as many of my audence as possible so I use a variety of posts each week and include all the subjects my audience likes. By using a variety of posts, I am reaching all my audience each week and I am posting only about the topics I know they want to see.

If you own a business that offers colorful yarns and classes on how to knit, then your audience includes people who are interested in crafts and how-to projects. These are people who are creative, with extra time to devote to hands-on tasks. Your audience also includes people with enough disposable income to afford what you have to sell. Your posts should include pictures of before and after projects, resources for classes, bundles of beautiful yarn, et cetera. If you create a meme or a video where a cat plays with yarn – you may double your chances for engagement.

If you want to begin a business that sells farm equipment, think of your audience. Your customers will be interested in the best bargains for high quality equipment. Farmers and ranchers want to know about various replacement parts for their equipment and how to obtain those parts quickly. They will also be interested in agricultural posts, best deals for buying land, how climate change is affecting our food supply, et cetera.

Of the examples listed above, each has its own niche and its own audience. The knitting business would not post about farm equipment and the farm equipment business would not post about yarn.

So to most effectively engage on social media, we must know our audience well. Then we must know the best type of posts for the most effective ROI – Return On Investment.

KISS – Keep it Short and Simple. Because most of us live busy lives, short and simple is best. We rarely take the time to read long posts unless we are following an instructional blog or learning from some type of online class.

The most effective posts are one-sentence quotes or memes which include a picture and a short line of text. The reader is instantly drawn to the picture and the text is easy to read.

Another effective post is the video – two minutes or less is preferable. These videos do not need to be professionally produced. In fact, some of the YouTube™ videos that go viral are initiated from a cell phone or a laptop.

An occasional post with just text is also effective, but again – keep it short and simple. This type of post might include a quote from a celebrity, an inspirational saying or a Bible verse. Easy to post. Easy to read. Good content but simple.

How do we decide which social media platform to use? How do we organize our time to effectively post and draw engagement?

Avoid the Overwhelm – unless your business hires an employee responsible for social media, you will not have time to learn every available platform. If you try to work with every platform every day, you can quickly become overwhelmed.

Again, knowing your audience is important. As a writer, most of my audience is on Facebook, Twitter and LinkedIn. So other than the blog posts I write, those are the three platforms I focus on for my social media engagement.

For the yarn store, Facebook, Pinterest and Instagram might be most effective. For the farm equipment business, Instagram and Facebook may work best.

Sometimes we have to try a platform for a while before we decide which one(s) to use. It is most effective to use at least two platforms, but not necessary to do more than three or four – unless you want to spend all your extra time on social media.

Once you know your audience, you can effectively learn how and when to post. My Facebook audience is available all day, but I find they are most ready to engage from 10am – 2pm, then in the evening from 6 – 8. So I use my lunch break and a few minutes in the evening to focus on Facebook.

The Twitter audience is a younger demographic, so they engage more often in the evenings. I also link my Tweets to Facebook so I draw double the engagement.

LinkedIn tends to be people in business, those who work the 8 - 5 shift and self-employed entrepreneurs. So I focus my LinkedIn posts on engagement during the work day with industry leaders and writing groups.

Once you decide on your social media platform(s), study the demographics and find out what time of day is best. As you post at peak times, you will affectively increase your ROI.

Although marketing on social media is important for a writer, I do not want to spend my entire day making memes, searching for effective quotes and engaging with my audience.

My mission is twofold: to write my projects and to coach my writing clients. So I only spend a total of 30 – 45 minutes each day on social media. I engage with my audience, but I avoid the overwhelm. I focus on my mission and spend most of my time either writing or coaching my clients.

Why does this fairly small amount of time work? Because of the key component necessary for social media engagement.

Be Consistent – Consistency is the key element for any type of engagement. We simply cannot build an audience or affect our customers if we are not consistently engaging. We cannot have a presence on any social media platform without consistency.

This key element means doing some type of post Every. Single. Day.

Every day? Oh, sure, it is impossible to post during illness, a weather disaster or if the internet is down. And I personally observe an electronic break on Sundays because it is important for me to commit to a day of rest. But for most of us, we can find at least 15 minutes each day to post something on our social media platforms.

Another way to remain consistent is to spend one day writing all your posts, then schedule them for the week or month. For some business owners, a scheduling program such as HootSuite™ or Buffer™ works well. For my writing and coaching mission, except for my blog posts which are scheduled several weeks out, I prefer a daily engagement with my audience.

Constant Content™, a platform for buying and selling articles and website content, reminds us: "*The biggest pain point in content marketing is producing engaging content, consistently.*"

But how can we, as busy authorpreneurs and business owners find a way to effectively and consistently en-

engage with our audience? How can we work smarter and not harder?

The answer follows. But first, let's discover the 5 most effective strategies for social media engagement. An easy way to remember these strategies is with the letters: **E, E, E, I, C.**

Emotion

Whenever we can trigger an emotional response in our audience, they will engage further with us and they will remember our posts for longer periods of time.

We all need to laugh, so humorous posts are effective. But be cautious! Effective humor does not include any type of slander against individuals or institutions. Although we may think it is funny to make fun of a politician or a particular demographic of people, that type of post will not increase quality engagement. In fact, it will cause distrust. And if we antagonize a segment of our audience, then our ROI will decrease.

If you have any doubt about a particular post, don't use it. This includes the practice of sharing another post. When in doubt, be cautious!

Writing and posting with effective humor is a creative gift, and I envy those who do it well. So I borrow from other humorous posts and share videos about pets doing funny things or children saying something hilarious. A favorite is any video with a laughing baby. If I can incorporate that laughter in a post, it will gain automatic engagement and often will be shared.

Use humorous quotes from celebrities. One of my favorites is by Mark Twain, *"Everybody talks about the weather, but nobody does anything about it."*

The other most effective trigger is the one that makes us cry or tear up a bit. Videos that depict the soldier dad or mom coming home to surprise the first grader at school. I always cry a bit while watching those videos and I always click on them – even though I know what my response will be.

A calmer type of emotional post is the one that makes us say, "*Aww – so sweet.*" Examples might be the mother cat who adopts the baby bunny, the big sister who kisses her infant brother or fuzzy ducklings waddling across the highway while traffic stops.

Fear is also an effective emotion. Even the possibility of dread may increase your engagement. One of my effective posts is about Alzheimer's Disease. "*Over 60 million people become caregivers of an Alzheimer's patient. Will that be you? How will you handle the 36 hour-day?*" Then I include the link for my book, "Sometimes They Forget." http://amzn.to/2uqi3mT

In this example, I also use the effectiveness of numbers. They are eye-catching, easy to read and understandable. Remember: short and simple.

So think about the types of posts you can use to add emotion. Whether you post humor, sorrow, fear or even dread – the addition of emotion can increase your engagement and help your audience remember what you have shared.

Education

For writers with an educational background, this type of social media engagement comes easy. We share facts or some type of interesting information with the intention of educating our audience. Life-long learners love these types of posts and may click onto your website for even more information.

One of the easiest ways to post an educational or informational topic is to use your research. If you have written a book – whether nonfiction or fiction – you have spent some time in research. Use those facts to engage with your followers.

For example: in writing about Alzheimer's, I learned about Turmeric and Rosemary. Both of these supplements come from plants and both are good for the brain. So I regularly include posts that educate my followers about the health of the brain. For example: "*Did you know that smelling a rosemary plant each day may increase cognitive brain health?"* A simple educational fact from my Alzheimer's research.

Another fact from my research involves processed sugar. We all know too much sugar can lead to cavities. But research is also beginning to show how sugar may grow cancer cells and possibly lead to cognitive impairment.

So I share those nutritional facts with my followers, hoping to interest them in Alzheimer's research. Then as my followers learn about dementia and Alzheimer's Disease, I anticipate they will buy my books.

Travel writers educate their followers by telling them about places to go and things to see in that location. Food bloggers share recipes and some of the advantages of the newer kitchen gadgets.

Amy Bovaird is a writer who lives with retinitis pigmentosa. Through her engaging posts, she taught me about RP and its effects on her life. To engage with Amy, you can follow her here. http://AmyBovaird.com/

If you write historical fiction, you will research about clothing, recreational activities, cultural stigmas, foods and a host of other topics. Share those facts with your followers. Educate them about a certain time period and you will interest them in your books.

Another way to engage and educate is through the use of the how-to video. YouTube™ is filled with self-help videos. How many of us have Googled™ a tutorial? We need to learn something so Google™ is the first place we go. Take advantage of tutorials and other videos to engage with your followers.

Make your own videos using the webcam on your computer. A video will educate your followers if you are authentic, honest and straight forward. Just keep it short and simple.

One of the most effective ways to educate with a video is through the use of flash cards. Simply design a statement based on your research, such as "7 Tips for Caregivers." Make flash cards of information with your printer or handwrite them. Then display the flash cards as you speak. Simple. Effective.

When it comes to increasing your engagement on social media, your followers want to know you are an expert in your field. The best way to underscore your expertise is to share what you know with your followers.

Educate your audience and they will thank you by increasing their comments and their shares.

Entertainment

We have already mentioned some of the most effective ways to entertain your audience: videos and humor. We all like to watch videos and we all love to laugh. By learning how to make simple videos and by sharing something funny, you will engage with your followers.

But think outside the box. Certain types of posts are almost always guaranteed to entertain. Marketing experts maintain the following principle: "*Puppies and children can sell anything.*"

Most small animals are engaging and entertaining. Examples might include: a picture of a cat reading your book, a chipmunk munching on your newest granola recipe or a fawn drinking from the pond you designed.

Utilize the proven tools of entertainment: a post from a celebrity, the playbill for a Broadway show you attended, a video of the Rockettes or your favorite sports team in action. Be creative and weave various forms of entertainment into your posts.

One of the tools I use each time I launch a book is the song "Tomorrow" from the Broadway show "Annie." As we grow closer to the date of my launch, I count off the days on Facebook with images of large numbers: 10, 9, 8, 7, et cetera.

Then the day before the launch, I post a video of Annie singing "Tomorrow." It has become one of my signature tools and a proven way of entertaining my followers.

One of the advantages of entertainment is that is diverts us from daily life. It provides an escape from stress, routine and even calamity. Comedians are most effective when times are tough. Bob Hope and his touring USO groups proved the effectiveness of humor during national crises. They helped service men and women divert their minds from war and conflict by using the powerful tool of entertainment.

Remember how important it was to watch the Yankees play a baseball game after the tragedy of 9/11? We all needed something to return us to normalcy, to remind us we were Americans and we would not be defeated. We desperately needed the entertainment of our national past-time – a baseball game. And when President George W. Bush threw out that first ball – the cheers reverberated around our nation.

Using entertainment to engage on social media helps your followers remove themselves from personal tragedy and reminds them life can return to normal. Entertainment will help them remember you and cement your brand.

So practice making videos. Look for pictures of puppies and children. Find something funny. Post a YouTube™ video of an appropriate song. Take pictures of a before and after project. Show us how you make your grandmother's beef stew. Make a video of your successful book launch.

Think outside the box and ask yourself, "*What entertains me?*" Then engage with your followers through the powerful tool of entertainment.

Inspiration

Inspirational posts are the ones that make us feel good. They inspire us to do something beneficial for mankind and/or they motivate us to move forward in life. They help our hearts feel a bit better in a scary world.

These are the posts your followers will watch for and appreciate when they are feeling discouraged. An inspirational post can impact your audience, and it will often be marked as a favorite then shared with others.

One of the most effective ways to use an inspirational post is with a meme – a picture with some type of text printed on it. For writers, effective memes can underscore a review and send the reader to your website for more books. Always include the url of your website somewhere on the meme, so your followers and new friends are reminded how to find you.

The safest photos to use are the ones you shoot yourself. Keep a file on your computer of your favorite social media pictures. You can use them over and over – just post a different text each time.

But if you do not have a photo that works for your particular post, check out some of the free sites that offer pictures. Be careful to read about the attribution clause. If a picture is copyrighted, then you should not use it.

Some of my favorite sites are: Pixabay, Canva and Pik Monkey. On Canva and Pik Monkey, I can design frames around my photos, color the background, use some interesting texture, et cetera. These sites provide a fun way to be creative while I focus on inspiring my audience.

You can also make one of your own statements into a meme. Use Canva or Pik Monkey to add the text box, the color, the texture and a particular font. I regularly use my own text from a blog post or a sentence from one of my books to make a meme. It can be as simple as: "*Shared sorrow expands hearts.*" Check out the blog post that includes this quote at: https://RJThesman.net/2017/09/05/hope-finds-a-miracle/

Another way to inspire your followers is through the use of inspirational quotes from other writers. Are you reading a book and you just experienced an a-ha moment? Quote that author on social media and inspire others. Always include the attribution. The following quote is an example I used this week, "*Grief remains one of the few things that has the power to silence us.*" – Anna Quindlen.

Do you have a favorite author? Research some of the inspirational quotes he or she has written through the years. Maya Angelou is always a favorite. So is Mark Twain, plus he is also entertaining. Other authors I often quote are Brene Brown, Anne Lamott and Richard Foster.

Some Bible verses can be inspirational, but be careful not to preach to your audience. Carefully introduce God's love rather than pummeling your audience with judgment and wrath. Our job is to inspire; not convict.

Another effective way to inspire your audience is through holidays or emphasizing certain national days. For example: during the month of November, I post about

Alzheimer's Disease because November is the month for learning about Alzheimer's awareness.

During each of the holidays, I change my cover photo on Facebook. For example: August 9 is National Book Lover's Day. I post pictures of books and libraries, ask my followers to write a review and hopefully – inspire them to read more of my books.

Be creative in ways you inspire using your books or your products. Does one of the characters in your book have an inspirational quote he repeats? Make a meme of that quote. Include the url of your book so readers can easily find it on Amazon™ or your website. Make it easy for readers to find you and purchase your products.

Writers impact the world through the use of words. We inspire young people to become writers. We share positive thoughts that help lift the discouraged. We post about love, joy and peace.

Be an inspirational engager on social media. Make the world a better place because of your inspirational posts.

Call to Action

The call to action is one of the most effective posts on social media, yet it can be the most challenging to master. Most writers have a difficult time with the call to action, probably because we tend to be introverts. We would rather not force ourselves or our opinions on anyone else.

However, at some point – we need to promote our work, our books and our products. You may have noticed how I have used the call to action in previous pages of this book to promote my website, my blog posts and my books.

The call to action is basically telling people to buy your product. It is an invitation, strongly-worded, and a form of persuasion that results in sales. It is shameless self-promotion.

Marketers underscore the fact that consumers must be reminded **seven times** before they will buy something. **Seven times.**

This is why pizza commercials run in the evenings when they know we are tired and we do not want to make dinner. The gooey cheese, slices of pepperoni, roasted green peppers and happy children eating with their families – all are marketing techniques that persuade us to jump up and order a pizza. Especially after we watch the same commercial several times.

An effective meme can be a call to action. The photo shows a heroine in her frilly early-1800's dress, her umbrella pointed to the gallant hero. The text reads, "*Will Penelope find true love? Order your book today.*"

"*Order your book*" is the call to action. It is telling the consumer what to do and giving information about how to buy the product.

We can be subtle, yet utilize the call to action. For example, I posted about "Intermission for Reverend G" – the second book in my Reverend G series: "*Find out what happens when Reverend G decides to get married.*"

"*Find out*" is a call to action. And *you* can discover more about Reverend G here. http://amzn.to/2uqqx1K. Did you catch that subtle call to action?

Each year on my birthday, I post a reminder: "*How can you help an author celebrate her birthday? By writing a review.*" Then I post the link to my latest book. "*Write the review*" is the call to action.

One of the easiest calls to action is to ask for followers on your author page. Once each month, I post on Facebook, "*Please connect with me. 'Like' my author page.*" A number of people always respond to this simple call to action.

I use the same technique on Twitter with my once per week tweet: "*Follow my #Hope series at https://RJThesman.net.*" This call to action has helped grow more followers for my blog.

Observe how other authors or business owners use the call to action. Then hone your skills for these obviously promotional posts. At least once each week, post a call to action. March through your fear of self-promotion. Then watch how your audience responds.

Charting Your Engagement

We have discussed the **5 Strategies for Social Media Engagement**: Emotion, Education, Entertainment, Inspiration, Call to Action.

How can we find a simple system to remember all the posts we need to make? When we feel "time poor" because we're so busy, how can we effectively utilize these 5 strategies?

Simple. Use an Excel chart or a Table with columns. At the beginning, it takes a bit of thinking. But you CAN do it.

In the left column, write the 5 terms: Emotion, Education, Entertainment, Inspiration, Call to Action.

In the middle column, write a post with its appropriate links and the name of the image you will include.

In the right column, mark the day of the week you will copy/paste that particular post.

You will repeat posts throughout the month, but that is okay. Remember, your followers need reminders at least **seven times**. At the end of the month, think about a new series of posts for the next four weeks. When you write a new book, promote it with new posts. Include the thumbnail image for each new book.

Below is an example of a table I use for my posts. I give you permission to use my table as a template. Of course, you can design your own and color code it for easy reference.

Emotion	"How will you take care of your aging parent? Tips for #Caregivers in "Sometimes They Forget."	M, Th
Education	"Over 30 million people in the U.S. suffer with #Alzheimers."	Tues, Sat
Entertainment	Video showing #Alzheimers patient responding to music.	Wed
Inspiration	"*Though my father and my mother forget me, the Lord will receive me.*" Psalm 27:10 NIV	Fri
Call to Action	Is today one of "those" days? Follow my #Hope series at: http://RJThesman.net	Tues

You can also increase your table or your Excel chart to include miscellaneous ideas for additional posts. Some examples might be: the Amazon™ urls for each of your books, the location for your Amazon™ author page, links for specific YouTube videos that deal with your topic, file locations for pictures that illustrate your brand, et cetera.

If you have written several books, you can make a chart or even a simple list to focus on one book for each day of the week. For example:

> "The Unraveling of Reverend G" – Monday –
> http://amzn.to/2eTjA2w
> "Intermission for Reverend G" – Tuesday –
> http://amzn.to/2xUkK2h
> "Final Grace for Reverend G" – Wednesday –
> http://amzn.to/2wOjnVi
> "Sometimes They Forget" – Thursday –
> http://amzn.to/2jbmiSr
> "Setting & Reaching Your Writing Goals" – Friday –
> http://amzn.to/2jatwWO

Did you notice my not-so-subtle **Call to Action** as I listed some of my books? When you use the same method and include the urls, you give your followers the information they need to easily purchase your products.

No rules apply on how you make your own table or chart. Keep it short and simple. Pull up your chart each day, copy / paste for that day – then watch the engagement happen.

As you reply on each comment, your engagement may also increase as well as interest in your next post.

Our followers want us to be available to them. Using an Excel chart, a Table with columns or a simple list helps keep social media engagement easy and avoids the overwhelm.

Accountability Questions

As a writing coach, I give my clients accountability questions to help them learn how to engage on social media platforms. Consider the following questions:

What one thing do you need to change to more strategically engage on social media?

What is a post you could write using the following topics?

Emotion:

Education:

Entertainment:

Inspiration:

Call to Action:

How will you become more consistent with your social media engagement?

What is the first action step you will take to improve your effectiveness on social media?

Summary

I hope this book has been helpful to you. Let me know if I can help you further by contacting me at: Rebecca@RJThesman.net.

Remember: your words are important and your product needs to be shared. The best way to promote your words is to learn more effective strategies to engage on social media and then be consistent with your posts.

So go forward to meet your action steps. Complete your Table or Excel chart and be committed to engage with your audience every day. Become the writer you were meant to be.

Disclaimer

Throughout this book, every effort has been made to accurately represent the **5 Strategies for Social Media Engagement**. However, there is no guarantee you will become an expert in social media, a best-selling author and/or make enormous amounts of money. Social media is always changing. This book does not guarantee a publishing contract, increased sales or multitudes of followers.

Examples in this book are not to be interpreted as a promise or guarantee of earnings. Everything depends on you and how committed you are to learning more about social media engagement and being committed to reach your goals.